I0468051

Resources for Evaluating and Monitoring Climate Change Adaptation Actions in Coastal Regions: An Annotated Bibliography

November 2013

United States
Global Change
Research Program

Adaptation Science Interagency Working Group

Table of Contents

Table of Contents 2

Introduction 4

Evaluation 5

Resources for Evaluation Practices and Approaches 11

Monitoring 20

Evaluation Organizations 30

Glossary 31

About the USGCRP Adaptation Science Interagency Working Group

In 2009, the Obama Administration convened an Interagency Climate Change Adaptation Task Force (ICCATF)[1] with the participation of more than 20 Federal agencies. Shortly thereafter, President Obama signed *Executive Order 13514: Federal Leadership in Environmental, Energy, and Economic Performance*,[2] directing the ICCATF to recommend ways the Federal Government can strengthen the Nation's ability to adapt to the impacts of climate change. Several Workgroups were established under the ICCATF including one on Adaptation Science. This Work Group was soon transferred to the U.S. Global Change Research Program (USGCRP)[3] in 2010 in recognition of USGCRP's mission for advancing global change science and making research products available and translated to support and inform adaptation actions. The mission of the USGCRP Adaptation Science Interagency Workgroup (ASIWG)[4] is to ensure that Federal science effectively informs adaptation decisions at a range of scales, in diverse sectors. It also provides scientific support to agencies in the adaptation planning process established under EO 13514. For more information on the USGCRP Adaptation Science efforts, see: http://globalchange.gov/what-we-do/prepare-the-nation-for-change.

This report is a product of the ASIWG, with the National Oceanic and Atmospheric Administration (NOAA) and the U.S. Army Corps of Engineers (USACE) taking a lead role in its preparation.

---

[1] White House Council on Environmental Quality. *Task Force Progress Reports*. Website: http://www.whitehouse.gov/administration/eop/ceq/initiatives/resilience
[2] White House. *Executive Order 13514: Federal Leadership in Environmental, Energy, and Economic Performance*. Federal Register Vol. 74, No. 194. October 2009. Website: http://www.gpo.gov/fdsys/pkg/FR-2009-10-08/pdf/E9-24518.pdf
[3] U.S. Global Change Research Program. *About*. Website: http://globalchange.gov/about
[4] U.S. Global Change Research Program. *Prepare the Nation for Change*. Website: http://globalchange.gov/what-we-do/assess-the-us-climate

Introduction

Federal agencies in the United States are planning for climate change in many ways, including through the design of climate change adaptation actions. Like any iterative planning process, climate adaptation planning includes monitoring and evaluation components so plans and actions can be measured for their overall effectiveness. It is important for program and project managers responsible for designing and implementing climate adaptation actions to be able to monitor and evaluate the success (or at times the failure) of their efforts. Learning from monitoring and evaluating the effectiveness of adaptation actions can help the Federal government adjust their actions and responses accordingly to strengthen the resilience of communities around the Nation.

To better support Federal agencies in their coastal climate adaptation planning efforts, the U.S. Global Change Research Program (USGCRP)'s Adaptation Science Interagency Working Group (ASIWG) conducted a search of numerous databases to identify examples of programs and products related to evaluations of coastal climate adaptation actions. The number of evaluations of climate adaptation actions was small, however, and those that had been done focused largely on the early stages of programs and projects. This is not surprising, as most organizations and governments are only beginning to implement climate adaptation planning and actions. To enlarge the set of evaluations for review, the focus was widened to include evaluations of other actions related to climate adaptation (e.g., those related to natural resource management, disaster risk reduction, etc.) and their evaluation resources. This review also explored the literature related to continued monitoring specifically for adaptation planning and actions, since observations and measurements are crucial for successful evaluations.

This annotated bibliography presents the results of that review of the relevant peer-reviewed literature and reports from Federal, state, and local governments. It includes journal articles and publicly available reports in addition to a number of guides, websites, tools, and other resources developed for program and project managers interested in planning and conducting evaluations. A glossary of common monitoring and evaluation terms is provided at the end of this document.

The ASIWG hopes that this bibliography will be a useful resource for program and project managers looking to inform their decisions on climate change adaptation planning and the evaluation programs that will help ensure their success.

Evaluation

This section provides examples of evaluations of climate adaptation plans and actions, as well as examples of evaluations of related actions. In addition, it includes a few examples of evaluations of climate adaptation monitoring and evaluation frameworks.

Examples of Evaluations of Climate Adaptation Actions

Committee on Climate Change (United Kingdom). Adaptation Sub-Committee. 2011. How Well is Scotland Preparing for Climate Change?
http://www.theccc.org.uk/publication/how-well-is-scotland-preparing-for-climate-change-asc-scotland-report-2011

The U.K. Adaptation Sub-Committee's first report to the Scottish government analyzes the Scottish government's adaptation framework against the subcommittee's preparedness ladder (see *How Well Prepared is the UK for Climate Change?* On p. 11) based on reviewing published adaptation plans and strategies to provide an initial assessment of how well Scotland is preparing for climate change. The report includes a review of physical, socioeconomic, and environmental data; provides a summary of Scotland's adaptation policy framework; analyzes progress in priority areas for early adaptation; and provides recommendations to further enable adaptation action by key private and public sector organizations, communities, and individuals.

Committee on Climate Change (United Kingdom). Adaptation Sub-Committee. 2012. Climate Change—Is the UK Preparing for Flooding and Water Scarcity?
http://www.theccc.org.uk/publication/climate-change-is-the-uk-preparing-for-flooding-and-water-scarcity-3rd-progress-report-2012

The U.K. Adaptation Sub-Committee's third progress report documents the use of the subcommittee's adaptation assessment toolkit (see the first and second reports on pp. 11 and 22, respectively) to assess progress toward preparing for flooding and water scarcity associated with climate change. The toolkit has two main components: monitoring changes in climate risks using indicators in three broad categories—indicators of risk, indicators of adaptation actions, and indicators of climate impact—and evaluating preparedness for future climate. Based on the analysis, the report lays out advisory recommendations for the development of the U.K. National Adaptation Programme.

COWI and International Institute for Environment. 2009. Joint External Evaluation: Operation of the Least Developed Countries Fund for Adaptation to Climate Change. Ministry of Foreign Affairs/DANIDA (Denmark). Evaluation Department.
http://um.dk/en/danida-en/results/eval/eval_reports/publicationdisplaypage/?publication ID=B42FCD1C-B94E-424C-BDF9-A571F1710743

This report presents the details of an evaluation of the operations of the United Nations Framework Convention on Climate Change's Least Developed Countries Fund (LDCF) in financing and promoting climate change adaptation. The fund supports least developed countries in their efforts to prepare and implement national adaptation programs of action. The main evaluation criteria included relevance, effectiveness, efficiency, sustainability and—to the extent possible—impact issues. It focused on processes related to the production of programs of action and preparation and approval of related priority projects. It also analyzed the relevance of LDCF related outputs and the possible catalytic effects of the work done by the LDCF in terms of increasing awareness of and action on adaptation to climate change.

Global Environment Facility. Evaluation Office. 2011. Evaluation of the Special Climate Change Fund.
http://www.thegef.org/gef/Program%20Evaluation%20-%20SCCF

This report presents an evaluation of the United Nations Framework Convention on Climate Change's Special Climate Change Fund (SCCF). The results of the evaluation illustrate the SCCF's progress toward its objectives and its main implementation achievements. The evaluation was structured according to four evaluation criteria: relevance, effectiveness, efficiency, and results and sustainability of results. Due to the early stage of implementation of most SCCF projects, it was not possible to systematically measure impact. However, examples of available results as well as potential sustainability of the results are included in the report.

Indian and Northern Affairs Canada. Audit and Evaluation Sector. 2009. Implementation Evaluation of INAC Climate Change Adaptation Program: Assist Northerners in Assessing Key Vulnerabilities and Opportunities.
http://www.aadnc-aandc.gc.ca/eng/1307024581995

This report provides the results of an implementation evaluation of Indian and Northern Affairs Canada's Assist Northerners in Assessing Key Vulnerabilities and Opportunities Program. The program provides funding to help communities in Northern Canada understand the impacts of climate change and take steps to respond to anticipated changes. The evaluation, guided by a matrix of questions, indicators, and data sources, focused on relevance of adaptation programming, the design and implementation of the program, and the program's preliminary results/success. The report describes the program, defines the evaluation scope, outlines the evaluation methodology, presents the findings, and offers recommendations, which include identifying and allocating specific resources to performance measurement.

Indian and Northern Affairs Canada. Audit and Evaluation Sector. 2011. Evaluation Update of the Climate Change Adaptation Program: Assist Northerners in Assessing Key Vulnerabilities and Opportunities Program.
http://www.aadnc-aandc.gc.ca/eng/1322226054441

This report provides an update to the first evaluation of Indian and Northern Affairs Canada's Assist Northerners in Assessing Key Vulnerabilities and Opportunities Program (see previous entry). The program provides funding to help communities in Northern Canada understand the impacts of climate change and take steps to respond to anticipated changes. The evaluation, guided by a matrix of indicators and data sources, focused on the program's achievement of outcomes and its economy, efficiency, and cost-effectiveness. The report describes the program, defines the evaluation scope, outlines the evaluation methodology, summarizes the findings of the previous evaluation, presents the current findings, and offers recommendations.

Natural Resources Canada. 2009. Evaluation of the Canadian Forest Service Climate Change Program.
http://www.nrcan.gc.ca/evaluation/reports/2009/2785

This report provides the findings of an evaluation of the Canadian Forest Service's Climate Change Program. The program aims to understand the climate change impact implications for Canada's forests and forest sector and develop adaptation strategies to reduce the risks and ensure that Canada's climate change international reporting obligations and commitments are met and that related greenhouse gas mitigation options are explored. The evaluation, which was based on a document review, structured interviews and discussions, and case studies, examined three issues: relevance and rationale, results and success, and cost-effectiveness and alternatives. The report provides background on the issue, describes the evaluation scope and methodology, presents the findings, and offers recommendations.

Natural Resources Canada. 2011. Evaluation of the Climate Change Geoscience and Adaptation Program Sub-Activity.
http://www.nrcan.gc.ca/evaluation/reports/2011/3642

This report provides the findings of an evaluation of Natural Resources Canada's Climate Change Geoscience and Adaptation Program Sub-Activity. The sub-activity aims to help Canadians understand and prepare for climate change by generating and incorporating new knowledge into planning and resource management. The evaluation, which was based on a document review, stakeholder interviews, and case studies, assessed the program's relevance and the extent to which it achieved expected outcomes and demonstrated efficiency and economy. The report describes the programs, defines the evaluation scope, outlines the evaluation methodology, presents the findings, and offers recommendations.

Lafontaine, A., J.O. Adejuwon, P.N. Dearden, and G. Quesne. 2012. Final Evaluation of the IDRC/DFID Climate Change Adaptation in Africa Programme. Canadian International Development Research Centre and U.K. Department for International Development.
http://r4d.dfid.gov.uk/Output/191377/Default.aspx

This report presents the evaluation of the Canadian International Development Research Centre and the U.K. Department for International Development's jointly funded Climate Change Adaptation in Africa (CCAA) research and capacity development program. The CCAA was created to improve the capacity of African populations and organizations to adapt to climate change in ways that benefit the most vulnerable. The evaluation focused on the extent to which the CCAA achieved its goal and objectives, the results of the program, the effectiveness of program management and governance, and identification of lessons learned. Data collection methods included interviews, focus groups, online surveys, documentation review, meetings with project partners, and project desk case studies. The report provides background on the program, describes the methodology, presents the findings, and offers recommendations.

World Bank. Independent Evaluation Group. 2013. Adapting to Climate Change: Assessing the World Bank Experience.
http://ieg.worldbankgroup.org/evaluations/adapting-climate-change-assessing-world-bank-group-experience

This is the third and final volume of a series of assessments of the World Bank Group's engagement with climate change, and the only volume to address adaption. The purpose of this evaluation was to draw from World Bank Group experiences with adaptation by answering questions related to dealing with climate variability, factoring climate change risks into investment projects, anticipating climate change, and the group's performance in meeting adaptation goals of the Strategic Framework on Development and Climate. The report concludes with a number of recommendations based on the findings to help the bank enhance its operational effectiveness on adaptation.

Examples of Evaluations of Preparedness and Resilience Actions

Brody, S.D., S. Zahran, W.E. Highfield, H.Grover, and A. Vedlitz. 2008. Identifying the impact of the built environment on flood damage in Texas. *Disasters* 32 (1): 1-18. $[5]$
http://dx.doi.org/10.1111/j.1467-7717.2007.01024.x

This journal article looks at the relationship between the built environment and flood impacts in the eastern portion of Texas. Researchers calculated property damage resulting from 423 flood events over a five-year period and identified the impact of several built environment variables (wetland alteration, impervious surface, and dams) on the damage while controlling for biophysical and socioeconomic variables.

Federal Emergency Management Agency. 2009. Loss Avoidance Study: Eastern Missouri, Building Acquisition. Part One: General Overview and Part Two: Detailed Methodology.
Part One: http://www.fema.gov/media-library/assets/documents/17998
Part Two: http://www.fema.gov/media-library/assets/documents/16520

This report features the results of a study of the effectiveness of residential acquisition/ demolition projects as flood mitigation in eastern Missouri. The study included a quantification of the losses avoided due to the implementation of the projects through analysis of storm events that occurred in 2008. It involved calculating the value of the losses that had been avoided by the implementation of the projects and comparing the losses avoided with the project costs. Part One contains a description of the general methodology and the results of the study. Part Two provides detailed documentation of the methodology and can be used as guidance for future loss avoidance studies specific to acquisition projects (see also *Loss Avoidance Study: A Handbook for Decision Makers* on p. 14).

Jones, C.P., W.L. Coulbourne, J. Marshall, and S.M. Rogers, Jr. 2006. Evaluation of the NFIP's Building Standards. American Institutes for Research.
http://www.fema.gov/library/viewRecord.do?id=2592

This report represents one of the 13 individual evaluations conducted for the evaluation of the National Flood Insurance Program (NFIP, see *The Evaluation of the National Flood Insurance Program—Final Report* on p. 8). Focused on the NFIP's building standards, it examines damage prevented or induced by strict adherence to minimum design and construction requirements and the costs and benefits of modifying the minimum requirements. Methods used included a literature review, meetings with NFIP staff and other experts, survey of floodplain professionals, a benefit-cost analysis, and use of an expert panel to guide the study and review its findings and recommendations.

Miller, T.R., E. Langston, and V. Nelkin. 2006. Performance Assessment and Evaluation Measures for Periodic Use by the National Flood Insurance Program. American Institutes for Research.
http://www.fema.gov/library/viewRecord.do?id=2575

This report represents one of the 13 individual evaluations conducted for the evaluation of the National Flood Insurance Program (NFIP, see *The Evaluation of the National Flood Insurance Program—Final Report* on p. 8). It evaluates the performance measures routinely used by the NFIP. Based on interviews of NFIP managers and stakeholders, this evaluation assessed the effectiveness of the performance measures for informing stakeholders and policymakers about the NFIP's costs, benefits, and accomplishments and for internal tracking of progress toward

---

[5] *$—The dollar sign here and throughout signifies that there is a cost associated with the resource. It is not available for free.*

goals. The report offers recommendations for new measures to improve performance assessment.

Monday, J., K. Grill, P. Esformes, M. Eng, T. Kinney, and M. Shapiro. 2006. An Evaluation of Compliance with the National Flood Insurance Program—Part A: Achieving Community Compliance. American Institutes for Research.
http://www.fema.gov/library/viewRecord.do?id=2589

This report represents one of the 13 individual evaluations conducted for the evaluation of the National Flood Insurance Program (NFIP, see *The Evaluation of the National Flood Insurance Program—Final Report* on p. 8). It evaluates community compliance with NFIP standards. Evaluation methods included open-ended interviews with select regional, state, and community floodplain management and insurance staff as well as specialists from the private and nonprofit sectors; shadowing staff on assistance visits conducted in four participating communities; and review of community files, policy guidance, and letter and communication templates as well as information from other associated reporting tools.

Sarmiento, C. and T.R. Miller. 2006. Costs and Consequences of Flooding and the Impact of the National Flood Insurance Program. American Institutes of Research.
http://www.fema.gov/library/viewRecord.do?id=2577

This report represents one of the 13 individual evaluations conducted for the evaluation of the National Flood Insurance Program (NFIP, see *The Evaluation of the National Flood Insurance Program—Final Report* on p. 8). It examines the cost effectiveness of the NFIP in reducing flood costs to residences in single-family housing areas as well as in reducing costs to taxpayers and how well the NFIP serves low-income households. The researchers ran simulations on HAZUS (loss estimation software) and used historical data on NFIP payouts and other federal and nonfederal flood-related compensation to estimate who (government or individuals) would pay for the predicted flood losses.

Sheridan, S.C. 2007. A survey of public perception and response to heat warnings across four North American cities: an evaluation of municipal effectiveness. *International Journal of Biometeorology* 52: 3-15. $
http://dx.doi.org/10.1007/s00484-006-0052-9

This journal article examines the effectiveness of municipal heat watch warning systems through an evaluation of the heat mitigation plans of four North American cities and a survey of residents in the metropolitan areas of these cities. The survey gauged residents' perceptions of their vulnerability to the heat and their knowledge of heat warnings and the activities that can help mitigate effects of the heat. It also sought to understand how residents reacted to a warning and why they did or did not take action.

State of Florida. Division of Emergency Management. 2012. Loss Avoidance Assessment: Tropical Storm Debby.
http://www.floridadisaster.org/Mitigation/SMF

This report features the results of an assessment of the effectiveness of flood mitigation projects in Florida funded through Federal Emergency Management Agency's Hazard Mitigation Assistance grant programs during four flood events, including Tropical Storms Fay (2008) and Debby (2012). It consists of two primary sections. Part I contains an introduction to loss avoidance assessments, project highlights and results, and lessons learned. Part II provides an outline of Florida's system and strategy to assess loss avoidance (see *Loss Avoidance Analysis System and Strategy Flood Mitigation—Building Modification Projects* on p. 15) and an explanation of how it was implemented for this report.

Wetmore, F. G. Bernstein, D. Conrad, C. DiVincenti, L. Larson, D. Plasencia, R. Riggs. 2006. The Evaluation of the National Flood Insurance Program—Final Report. American Institutes for Research.
http://www.fema.gov/national-flood-insurance-program/national-flood-insurance-program-evaluation

This report summarizes the results of a comprehensive evaluation of the National Flood Insurance Program (NFIP) to develop data and information needed to formulate better policies for floodplain management, risk assessment, and insurance and to support long-term planning and policymaking for the NFIP. It is based on 13 individual evaluations that, using widely varying methods (e.g., key informant interviews, engineering surveys, case studies, database analyses, flood hazard modeling, legal and regulatory reviews, and cost-benefit analyses), focused on a variety of subjects determined to be critical to assessment of the NFIP's progress. These individual evaluations are also available; a few are noted on the previous pages.

World Bank. Independent Evaluation Group. 2006. Hazards of Nature, Risks to Development: An IEG Evaluation of World Bank Assistance for Natural Disasters.
https://openknowledge.worldbank.org/handle/10986/7001

This report provides the results of an analysis of World Bank efforts to address natural hazards conducted to determine the relevancy of the bank's policy goals, its effectiveness in addressing the objectives of bank policy, effectiveness in building in-country institutional capacity, and success in incorporating lessons learned into bank operations. The analysis drew heavily on completed and ongoing independent and self-evaluations and relied on a number of evaluation techniques, including a desk review of the bank's project portfolio, a literature review, a panel of experts to advise the study team, surveys, and interviews. The report presents the findings and recommendations to strengthen the bank's disaster readiness. The methodology is included as an appendix.

Examples of Evaluations of Climate Adaptation Monitoring and Evaluation Frameworks

Hedger, M.M., T. Mitchell, J. Leavy, M. Greeley, A. Downie, and L. Horrocks. 2008. Desk Review Evaluation of Adaptation to Climate Change from a Development Perspective. Institute of Development Studies.
http://preventionweb.net/go/7845

This report presents an overview of approaches relevant to or used for the evaluation of climate change adaptation interventions and identifies main gaps in such evaluations. It discusses why evaluations of climate change adaptation interventions are needed, the key issues involved in evaluating adaptation interventions, approaches, and methods that have been used at different levels, and a framework for evaluating adaptation interventions.

Lamhauge, N., E. Lanzi, and S. Agrawala. 2012. Monitoring and Evaluation for Adaptation: Lessons from Development Cooperation Agencies. OECD Environment Working Papers, No. 38. OECD Publishing.
http://dx.doi.org/10.1787/5kg20mj6c2bw-en

This working paper assesses monitoring and evaluation frameworks used by development cooperation agencies for projects and programs with adaptation-specific or adaptation-related components. It discusses monitoring and evaluation approaches used and indicators, baselines, and targets for monitoring and evaluation for adaptation with a focus on five broad categories of adaptation to climate change: risk reduction; policy and administrative management; education, training, and awareness; research; and coordination. The paper concludes with lessons learned—examples of best practices that should help to inform future work in this area—and approaches to consider.

United Nations Framework Convention on Climate Change. 2010. Synthesis Report on Efforts Undertaken to Monitor and Evaluate the Implementation of Adaptation Projects, Policies, and Programmes and the Costs and Effectiveness of Completed Projects, Policies and Programmes, and Views on Lessons Learned, Good Practices, Gaps and Needs.
http://preventionweb.net/go/13697

This document synthesizes information on efforts to monitor and evaluate the implementation of adaptation projects, policies, and programs and the costs and effectiveness of these activities. It also introduces relevant approaches and concepts associated with monitoring and evaluation and the development and use of adaptation indicators and summarizes lessons learned, good practices, and gaps and needs.

Resources for Evaluation Practices and Approaches

This section reviews examples of guides and other resources developed to introduce program and project managers to monitoring and evaluation. Some of these resources are specific to climate adaptation actions while others are more general but still related to climate adaptation.

Guides for Evaluation of Climate Adaptation Actions

Ayers, J., S. Anderson, S. Pradhan, and T. Rossing. 2012. Partic
ipatory Monitoring, Evaluation, Reflection and Learning for Community-Based Adaptation: A Manual for Local Practitioners. CARE International.
http://www.careclimatechange.org/files/adaptation/CARE_PMERL_Manual_2012.pdf

This guide presents a framework for participatory monitoring and evaluation to support adaptive decision-making in vulnerable communities. It includes key concepts; discussion of the steps to take when designing a participatory monitoring, evaluation, reflection, and learning process; and a description of tools that could be used when designing the process. One of the authors presented on the guide on a webinar, which is available at http://www.seachangecop.org/node/1859.

Brooks, N., S. Anderson, J. Ayers, I. Burton, and I. Tellam. 2011. Tracking Adaptation and Measuring Development, IIED Climate Change. Climate Change Working Paper No. 1. International Institute for Environment and Development.
http://pubs.iied.org/10031IIED.html

This paper from the International Institute for Environment and Development presents a framework for climate change adaptation programming, including potential indicators, or indicator categories/types, for tracking and evaluating the success of adaptation support and interventions. It begins with a discussion of some of the key issues related to the evaluation of adaptation and outlines some of the main difficulties and constraints with respect to the development of adaptation indicators. Next, it proposes the evaluation framework and identifies indicator categories. Lastly, it offers key conclusions and outlines a theory of change that shows how use of the framework could lead to more effective adaptation investments for climate-resilient development.

Committee on Climate Change (United Kingdom). Adaptation Sub-Committee. 2010. How Well Prepared is the UK for Climate Change?
http://www.theccc.org.uk/publication/how-well-prepared-is-the-uk-for-climate-change

The U.K. Adaptation Sub-Committee's first progress report introduces a framework to measure, evaluate, and monitor how well the United Kingdom, primarily the central government, is preparing for climate change. The framework comprises three elements—desired outcomes from adapting, a ladder of key activities in delivering adaptation outcomes, and policy to enable delivery. The report uses the framework to begin to assess progress in adapting to climate change, drawing predominantly on departmental adaptation plans, in five priority areas: land use planning, providing national infrastructure, designing and renovating buildings, managing natural resources, and emergency planning.

Pringle, P. 2011. AdaptME: Adaptation Monitoring and Evaluation. UKCIP.
http://www.ukcip.org.uk/adaptme-toolkit

This toolkit was developed to support program and project managers in assessing their work and thinking through factors that make evaluation of climate change adaptation actions challenging. It provides guidance on the development of logic models, the process of identifying indicators and performance measures, identification of questions an evaluation of a program or project should answer, and choosing the type of evaluation to conduct.

Prowse, M. and B. Snilstveit. 2010. Impact Evaluation and Interventions to Address Climate Change: A Scoping Study. International Initiative for Impact Evaluation.
http://www.3ieimpact.org/media/filer/2012/05/07/Working_Paper_7.pdf

This working paper promotes the use of impact evaluation to ensure effective allocation of resources in addressing climate change. It provides a brief introduction to impact evaluation, summarizes how it has been applied to climate change and related environmental interventions in developing countries, discusses the limits and opportunities for conducting rigorous impact evaluations of climate change interventions, focuses on some of the key areas relevant for mitigation and adaptation interventions, and suggests ways in which impact evaluations could be implemented, using evaluations in a variety of policy fields (agriculture, water resource management, social protection, and disaster risk reduction) as examples.

Somida, J., A. Faye, and H. N'Djafa Ouaga. 2011. Toolkit for Planning, Monitoring and Evaluation of Climate Change Adaptive Capacities: Handbook and User Guide. AGRHYMET Regional Centre.
http://cmsdata.iucn.org/downloads/top_secac_agrhymet_english.pdf

This publication brings together and harmonizes existing project planning, monitoring, and evaluation methods and tools for climate adaptation actions into a toolkit for implementation in Africa. It presents tools for understanding key concepts, assessing vulnerability and adaptive capacity, prioritizing and planning adaptive actions, and conducting monitoring and evaluation of adaptive actions.

Spearman, M. and H. McGray. 2011. Making Adaptation Count: Concepts and Options for Monitoring and Evaluation of Climate Change Adaptation. Deutsche Gesellschaft für Internationale Zusammenarbeit.
http://www.wri.org/publication/making-adaptation-count

This paper provides adaptation and development practitioners with a framework for designing project-level monitoring and evaluation systems to track the success and failure of adaptation initiatives in the development context. It discusses the role of monitoring and evaluation in adaptation and lessons learned from early adaptation efforts and provides a six-step process, along with examples and resources, for developing adaptation-relevant monitoring and evaluation systems for use in developing countries. The paper's key framing principles are a focus on learning; results-based management; and the understanding that adaptation is a long-term process that requires flexibility.

U.S. Agency for International Development. 2009. Adapting to Coastal Climate Change: A Guidebook for Development Planners.
http://pdf.usaid.gov/pdf_docs/PNADO614.pdf

This guidebook is intended to help U.S. Agency for International Development staff and their development partners understand the climate change impacts expected to affect the coastal zone throughout the developing world and adaptation measures available to address these impacts. It considers the design and implementation of adaptation measures around a suite of coastal resource management goals. A short chapter on evaluation focuses on evaluating for adaptive management.

Van den Berg, R. and O. Feinstein. 2010. Evaluating Climate Change and Development (World Bank Series on Development). Transaction Publishers. $

This book is a compilation of papers presented at the 2008 International Conference on Evaluating Climate Change and Development. It addresses climate change, development, and evaluation; provides a set of approaches and techniques for the monitoring and evaluation of climate change; and discusses challenges and lessons learned.

Evaluation Guides for Sectors or Climate Impact

Association of Fish and Wildlife Agencies. Teaming with Wildlife Committee. 2011. Measuring the Effectiveness of State Wildlife Grants.
http://teaming.com/tool/measuring-effectiveness-state-wildlife-grants-final-report-2011

This report recommends a framework that states, tribes, and their partners can use to assess the effectiveness of conservation actions funded through the U.S. Fish and Wildlife Service's State and Tribal Wildlife Program. The framework includes a list of generic conservation actions and a process for developing results chains, effectiveness measures for each action, and data collection questionnaires. The report also describes how the framework can be extended to assess the effectiveness of state wildlife action plans.

BetterEvaluation. n.d. BetterEvaluation. Website. Accessed June 21, 2013.
http://betterevaluation.org

BetterEvaluation is an international collaboration to improve evaluation theory and practice by sharing information on evaluation options and approaches. It has created a website to guide evaluators and program and project managers through the rapidly expanding range of choices available when planning and designing evaluation activities. The website divides evaluation into seven steps—manage, define, frame, describe, understand causes, synthesize, report, and support—and provides options and useful resources for each step of the process. In addition, it provides a forum for discussion and hosts articles on a variety of evaluation issues.

Centers for Disease Control and Prevention. National Center for Environmental Health 2010. Learning & Growing through Evaluation: State Asthma Program Evaluation Guide.
http://www.cdc.gov/asthma/program_eval/guide.htm

This guide was written for state asthma programs, but it provides a useful framework for developing and implementing any evaluation plan. It is broken into three modules: The first module covers the evaluation planning process and provides plan templates. Module 2 provides guidance, tips, and tools for implementing evaluations. And, Module 3 applies these tools to the evaluation of program partnerships.

Centers for Disease Control and Prevention. Office of the Director. 2011. Introduction to Program Evaluation for Public Health Programs: A Self-Study Guide.
http://www.cdc.gov/eval/guide

This guide for public health programs is based on the Centers for Disease Control and Prevention's *Framework for Program Evaluation in Public Health* (see http://www.cdc.gov/eval/framework). The guide provides a process for evaluation based on six steps: engage stakeholders, describe the program, focus the evaluation design, gather credible evidence, justify conclusions, and ensure use of evaluation findings and share lessons learned.

Dozois, E., M. Langlois, and N. Blanchet-Cohen. 2010. DE: 201 A—Practitioner's Guide to Developmental Evaluation. J.W. McConnell Family Foundation and the International Institute for Child Rights and Development.
http://www.mcconnellfoundation.ca/en/resources/publication/de-201-a-practitioner-s-guide-to-developmental-evaluation

This guide discusses the purpose of developmental evaluation, when it is most appropriate, the competencies needed to be an effective developmental evaluator, and how-to conduct developmental evaluation. It also explores some of the challenges and issues associated with the practice.

Federal Emergency Management Agency. 2009. Loss Avoidance Study: A Handbook for Decision Makers.
http://www.calema.ca.gov/HazardMitigation/Documents/FEMA%20LAS%20Decision%20Makers%20Handbook%2012-2009.pdf

This handbook describes the Federal Emergency Management Agency's methodology for determining the cost-effectiveness of hazard mitigation projects based on actual hazard events. The methodology is based on numerous studies of flood mitigation projects and the adaptation of loss avoidance concepts to mitigation projects associated with other types of hazards. The handbook includes an overview of the methodology's application for riverine flood, earthquake, tornado, and wildfire loss avoidance studies; the type, condition, and level of detail of data required;

types of analyses that must be performed and the general procedures for conducting them; and a description of the losses that can be analyzed to assess project effectiveness. It also supports long-range planning for conducting studies.

See *Loss Avoidance Study: Eastern Missouri, Building Acquisition* on p. 6 for an example of this handbook's application. InterAction. 2012. Impact Evaluation Guidance Note and Webinar Series.
http://www.interaction.org/impact-evaluation-notes

This four-part series on impact evaluation was developed to build the capacity of nongovernmental organizations (and others) to demonstrate effectiveness by increasing their understanding of and ability to conduct high quality impact evaluations. *Guidance Note 1: Introduction to Impact Evaluation* (P. J. Rogers) provides an overview of impact evaluation and different methods, approaches, and designs that can be used for the different aspects of impact evaluation. *Guidance Note 2: Linking Monitoring & Evaluation to Impact Evaluation* (B. Perrin) focuses on how routine monitoring and evaluation activities can support meaningful and valid impact evaluation. *Guidance Note 3: Introduction to Mixed Methods in Impact Evaluation* (M. Bamberger) explains mixed methods evaluation design and potential applications and benefits. *Guidance Note 4: Use of Impact Evaluation Results* (D. Bonbright) discusses three areas crucial for effective utilization of evaluation results.

Leeuw, F. and J. Vaessen. 2009. Impact Evaluations and Development: NONIE Guidance on Impact Evaluation. Network of Networks for Impact Evaluation (NONIE).
http://siteresources.worldbank.org/EXTOED/Resources/nonie_guidance.pdf

This document provides guidance on evaluating the impact of a program or project. Developed for international development programs and projects, it is structured around nine key issues in impact evaluation: identify the (type and scope of the) intervention, agree on what is valued, carefully articulate the theories linking interventions to outcomes, address the attribution problem, use a mixed-methods approach, build on existing knowledge relevant to the impact of interventions, determine if an impact evaluation is feasible and worth the cost, start collecting the data early, and front-end planning is important.

National Oceanic and Atmospheric Administration. Coastal Services Center. 2012. Planning for Meaningful Evaluation.
http://www.csc.noaa.gov/digitalcoast/publications/meaningful-evaluation

This publication from the NOAA Coastal Services Center (CSC) summarizes a process for planning for a program or project evaluation, and it illustrates many key concepts using examples from the coastal management sector. Topics covered include determining an evaluation question, creating effective performance measures, and designing data collection and analysis. This publication is directly linked to the CSC training course of the same name.

National Science Foundation. Division of Research, Evaluation, and Communication. 2002. The 2002 User-Friendly Handbook for Project Evaluation.
http://www.nsf.gov/pubs/2002/nsf02057/nsf02057.pdf

This National Science Foundation (NSF) handbook was developed for NSF managers to use in the evaluation of NSF's educational programs. It includes a step-by-step process for conducting an evaluation, discussion of quantitative and qualitative evaluation data collection methods, and strategies that address culturally responsive evaluation.

State of Florida. Division of Emergency Management. 2012. Loss Avoidance Analysis System and Strategy Flood Mitigation—Building Modification Projects.
http://www.floridadisaster.org/Mitigation/SMF/documents/LA-SystemStrategy-BldgMod.pdf

This document describes the State of Florida's system and strategy for assessing the effectiveness of certain flood mitigation projects. Specifically, it contains the state's methodology for conducting loss avoidance assessments and for using Florida's Loss Avoidance Calculator for certain projects that mitigate flood hazards (acquisition, elevation, floodproofing, mitigation reconstruction). The calculator, available at http://www.floridadisaster.org/Mitigation/SMF, was built in Microsoft Excel because it can be quickly and easily adapted, stored, and transferred to other users. *See Loss Avoidance Assessment: Tropical Storm Debby* on p. 8 for an example of how this methodology has been applied.

Twersky, F. and K. Lindblom. 2012. Evaluation Principles and Practices: An Internal Working Paper. The William and Flora Hewlett Foundation.
http://www.hewlett.org/library/hewlett-foundation-publication/evaluation-principles-and-practices

This working paper from the William and Flora Hewlett Foundation focuses on advancing evaluation practices within the organization and is shared as part of an ongoing collective dialog about evaluation practice. It lays out evaluation principles, discusses organizational roles, advises program managers on how to incorporate evaluation into their programs to maximize the value of evaluations, and discusses special evaluation cases.

U.S. Environmental Protection Agency. National Center for Environmental Innovation. 2009. Guidelines for Evaluating an EPA Partnership Program (Interim).
http://www.epa.gov/air/caaac/guidelines/Guide_interim.pdf

These guidelines were developed for conducting evaluations of U.S. Environmental Protection Agency Partnership Programs (programs designed to proactively target and motivate external parties to take specific, voluntary environmental actions). The publication is intended to introduce the novice to program evaluation and walks the user through a seven-step framework for how to design and conduct an evaluation. These steps are plan for an evaluation, identify key stakeholders, update or develop the program logic model, develop evaluation questions, select the evaluation design, implement the evaluation, and communicate the evaluation results.

U.S. Environmental Protection Agency. National Estuary Program. 2007. National Estuary Program Evaluation Guidance.
http://water.epa.gov/type/oceb/nep/upload/2009_03_26_estuaries_pdf_final_guidance_sept28.pdf

This guidance document describes the process for evaluating the U.S. Environmental Protection Agency's 28 National Estuary Programs (NEP). The primary purpose of the evaluations is to help the agency determine whether the programs are making adequate progress implementing their comprehensive conservation and management plans. The guidance uses a logic model framework designed to help guide reporting on stages of progress toward restoring and maintaining the ecological integrity of each NEP. The logic model shows causal links among activities, partnerships, outputs, pressures, and short-term, intermediate, and long-term outcomes.

U.S. Government Accountability Office. 2012. Designing Evaluations. GAO-12-208G.
http://www.gao.gov/products/GAO-12-208G

This methodology transfer paper from the U. S. Government Accountability Office addresses the logic of program evaluation designs. It introduces key issues in planning evaluation studies of federal programs to best meet decision makers' needs while accounting for the constraints evaluators face. The guidance describes different types of evaluations for answering varied questions, the process for designing evaluation studies, and key issues to consider to ensure overall study quality. The guidance is particularly useful for new evaluators and consumers of evaluations.

U.S. Government Accountability Office. 2012. Designing Evaluations. GAO-12-208G.
http://www.gao.gov/products/GAO-12-208G

This methodology transfer paper from the U. S. Government Accountability Office addresses the logic of program evaluation designs. It introduces key issues in planning evaluation studies of federal programs to best meet decision makers' needs while accounting for the constraints evaluators face. The guidance describes different types of evaluations for answering varied questions, the process for designing evaluation studies, and key issues to consider to ensure overall study quality. The guidance is particularly useful for new evaluators and consumers of evaluations.

W.K. Kellogg Foundation. 2004. W.K. Kellogg Foundation Evaluation Handbook.
http://www.wkkf.org/knowledge-center/resources/2010/W-K-Kellogg-Foundation-Evaluation-Handbook.aspx

The W. K. Kellogg Foundation developed this handbook to guide evaluation of the human services projects funded by the foundation. The handbook considers evaluation to be a management and learning tool that can be used for more than documenting program impacts, noting that evaluation can also be used to improve program effectiveness, provide learning opportunities, and gain better knowledge about what works. The handbook provides a framework for thinking about evaluation in this manner and outlines a blueprint for designing and conducting evaluations and using evaluation results.

Resources to Conduct Evaluation

EVALUATION CHECKLISTS

Western Michigan University. The Evaluation Center. n.d. Evaluation Checklists. Website. Accessed July 17, 2013.
http://www.wmich.edu/evalctr/checklists

This website provides a number of checklists to assist with designing and developing evaluations. These checklists cover evaluation design and management, evaluation models, evaluation values and criteria, meta evaluation, and building and institutionalizing evaluation capacity.

DATA COLLECTION METHODS

National Oceanic and Atmospheric Administration. Coastal Services Center. 2007. Introduction to Survey Design and Delivery.
http://www.csc.noaa.gov/digitalcoast/publications/survey-design

This publication provides insight into the various types and methods of survey research, including sampling techniques, survey delivery mechanisms, effective survey questions, and effective reporting of survey results.

National Oceanic and Atmospheric Administration. Coastal Services Center. 2009. Introduction to Conducting Focus Groups.
http://www.csc.noaa.gov/digitalcoast/publications/focus-groups

This publication is an introduction to key elements and practices that can increase the success of a focus group effort. Topics include determining if a focus group is the right tool for a project, developing effective focus group questions, planning for a productive focus group, and analyzing the data.

Scheuren, F., ed. 2004. What is a Survey? 2nd edition.
https://www.whatisasurvey.info

This booklet was written for nonspecialists and was designed to promote a better understanding of what is involved in carrying out sample surveys. It covers what a survey is, planning for a survey, collection of survey data, survey quality, focus groups, designing a questionnaire, conducting pretesting, mail surveys, telephone surveys, and margin of error.

University of Idaho. Park Studies Unit. n.d. Using Focus Groups to Evaluate Park Programs, Activities, and Visitor Services: A Training Program for Natural and Cultural Resource Services. Website. Accessed July 17, 2013.
http://psu.uidaho.edu/focusgroup

This online training program was developed to help natural and cultural resource professionals associated with the U.S. National Park Service education and cultural programs use focus groups to evaluate their activities, services, and programs. The program provides background information on focus groups, a how-to manual for conducting focus groups, and a how-to manual for group moderators.

DEVELOPING EVALUATION QUESTIONS

Preskill, H. and N. Jones. 2009. A Practical Guide for Engaging Stakeholders in Developing Evaluation Questions. Robert Wood Johnson Foundation.
http://www.rwjf.org/en/research-publications/find-rwjf-research/2009/12/the-robert-wood-johnson-foundation-evaluation-series-guidance-fo/a-practical-guide-for-engaging-stakeholders-in-developing-evalua.html

This guide focuses on engaging stakeholders in developing an evaluation's key questions. It lays out a five step process: prepare for stakeholder engagement, identify potential stakeholders, prioritize the list of stakeholders, consider potential stakeholders' motivations for participating, and select a stakeholder engagement strategy. The authors presented on the guide in a webinar, which is available at
http://www.fsg.org/tabid/191/ArticleId/383/Default.aspx?srpush=true

LOGIC MODELS

University of Wisconsin Extension. n.d. Enhancing Program Performance with Logic Models. Accessed July 15, 2013.
http://www.uwex.edu/ces/pdande/evaluation/evallogicmodel.html

The University of Wisconsin Extension developed this interactive online course for logic model development for faculty, staff, and partners to help them understand and use the logic model in planning and evaluating education and outreach programs.

W.K. Kellogg Foundation. 2004. Logic Model Development Guide.
http://www.wkkf.org/knowledge-center/resources/2006/02/WK-Kellogg-Foundation-Logic-Model-Development-Guide.aspx

The W.K. Kellogg Foundation developed this guide as a companion to the *W.K. Kellogg Foundation Evaluation Handbook* (see p. 17). The guide provides practical in-depth guidance on the principles of logic models, how to develop program logic models, and how to use a logic model to plan for evaluation.

VALUATION TECHNIQUES

Natural Capital Project. n.d. Integrated Valuation of Environmental Services and Tradeoffs. Accessed August 19, 2013.
http://www.naturalcapitalproject.org/InVEST.html

This suite of free software models was developed by the Natural Capital Project. The InVEST (Integrated Valuation of Environmental Services and Tradeoffs) software can be used by decision makers to map and compare terrestrial, freshwater, and marine ecosystem services over a range of scenarios. Depending on the data available, models can be validated and provide useful estimates of the magnitude and value of services provided.

National Research Council. Committee on the Effects of the Deepwater Horizon Mississippi Canyon-252 Oil Spill on Ecosystem Services in the Gulf of Mexico. 2013. An Ecosystem Services Approach to Assessing the Impacts of the Deepwater Horizon Oil Spill in the Gulf of Mexico. The National Academies Press.
http://www.nap.edu/catalog.php?record_id=18387

This publication reports on the results of an evaluation of the effects of the Deep Water Horizon oil spill on the ecosystem services of the Gulf of Mexico. It presents an ecosystem services approach to damage assessment and restoration that focuses on the valuable goods and services the resources provide (rather than the resources themselves) and discusses the potential benefits and limits of taking such an approach. It also examines the concept of resilience in the context of ecosystem services, highlights four case studies that demonstrate how an ecosystem services approach might be applied under various conditions and across wide levels of understanding of the services, and identifies gaps in existing data and models and research needs.

National Research Council. Committee on Assessing and Valuing the Services of Aquatic and Related Terrestrial Ecosystems. 2004. Valuing Ecosystem Services: Toward Better Environmental Decision-Making. The National Academies Press.
http://www.nap.edu/catalog.php?record_id=11139

This report is based on an evaluation of methods for assessing the economic value of the goods and services provided by aquatic and related terrestrial ecosystems to human societies. It outlines the major nonmarket methods currently available for estimating economic (monetary) values for these goods and services, discusses the applicability of each method, and includes case studies that show how they have been applied. It also discusses the use of professional judgment when conducting such studies and presents a set of guidelines for conducting or evaluating them. This report does not provide instructions on how to apply the methods, but provides a listing of references that can be used to develop a greater understanding of the methods.

U.S. Global Change Research Program. 2011. Valuation Techniques and Metrics for Climate Change Impacts, Adaptation, and Mitigation Options: Methodological Perspectives for the National Climate Assessment. NCA Report Series, Volume 8.
http://library.globalchange.gov/national-climate-assessment-valuation-techniques-and-metrics-workshop-report

This report summarizes the discussions and outcomes of a workshop focused on techniques for quantitatively valuing climate impacts and adaptation conducted to support the development of the 2013 National Climate Assessment (NCA). The goal of the workshop was to provide a snapshot of the capabilities, readiness, and applicability of methodologies for quantitatively valuing (i.e., placing a monetary or other type of value on) climate impacts and adaptation. Three primary topics were explored during the workshop: the current valuation landscape and the role of economic and noneconomic techniques in that landscape, general principles for applying valuation techniques for climate change analysis in a consistent and comparable manner, and the nature, applicability, and boundaries of economic techniques for valuation.

Monitoring

This section provides examples of monitoring climate adaptation plans and actions, as well as examples of sector-specific indicators. In addition, it includes a few examples of report cards and performance scorecards to monitor progress.

Examples of Multi-sector Indicators

AEA Technology Environment, Stockholm Environment Institute, and Metroeconomica. 2005. Objective Setting for Climate Change Adaptation Policy. U.K. Department for the Environment, Food, and Rural Affairs.
http://www.ukcip.org.uk/wordpress/wp-content/PDFs/Objective_setting.pdf

The U.K. Department for the Environment, Food, and Rural Affairs commissioned a project to develop a set of potential objectives, targets, and indicators for adaptation to climate change. This report describes the methodology for adaptation policy setting developed by the project team. It also includes strawman policy objectives for six sectors (water resources, flood and coastal risk management, transport, tourism, agriculture, and energy) developed using a risk-based approach, proposes examples of targets for some of the objectives, and identifies a number of existing indicators that are relevant to adaptation.

California Environmental Protection Agency. Office of Environmental Health Hazard Assessment. 2013. Indicators of Climate Change in California.
http://oehha.ca.gov/multimedia/epic/2013EnvIndicatorReport.html

This report from the California Environmental Protection Agency presents a compilation of environmental indicators that collectively describe changes to California's climate, the drivers of these changes, the impacts of such changes on the state's physical and biological systems, and emerging climate change issues. The indicators draw upon data collection, monitoring, and studies by state and federal agencies, universities, and research institutions. Information for each indicator includes trend data, a discussion of why the indicator is important, the factors that influence the indicator, and information on the data characteristics (including strengths and limits).

CARE International. 2010. Framework of Milestones and Indicators for Community-Based Adaptation. http://www.careclimatechange.org/tk/integration/en/quick_links/tools/monitoring_evaluation.html

This document is a working draft of milestones and indicators focused on monitoring and evaluating adaptive capacity rather than fixed outcomes. It includes a set of proposed milestones and indicators to help plan activities and track progress toward achieving "enabling factors," which must be in place at different levels for effective community-based adaptation to happen. The proposed milestones, indicators, and indicator definitions are provided at the household/individual, local government, and national levels and are grouped by enabling factors under four strategies: climate-resilient livelihoods, disaster risk reduction, capacity development, and addressing underlying causes of vulnerability.

Committee on Climate Change (United Kingdom). Adaptation Sub-Committee. 2011. Adapting to Climate Change in the UK: Measuring Progress. http://www.theccc.org.uk/publication/adapting-to-climate-change-in-the-uk-measuring-progress-2nd-progress-report-2011

The U.K. Adaptation Sub-Committee's second progress report builds on previous reports and describes further development of the framework to monitor preparedness for climate change. The report sets out a range of indicators against which progress will be measured and focuses on three priority areas: land use planning, managing water resources, and the design and renovation of buildings.

Harvey, A., N. Hodgson, M. Benzie, S. Winne, R. Smithers, S. Dresner, P. Drummond, C. Coleman, L. Horrocks, and M. Harley. 2011. Provision of Research to Identify Indicators for the Adaptation Sub-Committee. AEA Technology. http://archive.theccc.org.uk/aws2/ASC%202nd%20Report/ED56687%20Final%20Report%20Issue%203_130711.pdf

This report documents the results of a project conducted in support of the U.K. Adaptation Sub-Committee's work to establish a framework for monitoring preparedness for climate change (see previous entry). It describes the conceptual framework used to develop indicators of impacts, drivers of those impacts, and adaptation actions and applies the framework to five broad sectors: health and well-being, built environment, critical infrastructure (water, energy, and transport), natural environment, and business and the economy. The report also contains information on data sources and recommendations for future work to improve the indicators.

National Research Council. Committee on Indicators for Understanding Global Climate Change. 2010. Monitoring Climate Change Impacts: Metrics at the Intersection of the Human and Earth Systems. The National Academies Press. http://www.nap.edu/catalog.php?record_id=12965

This report is the result of work to develop a suite of indicators, measurements, and metrics that are most important for understanding global climate change and how it will affect the five domains of human vulnerability (food, water, energy, shelter, and health). Written for analysts in the intelligence community, as well as researchers, as they delve more deeply into climate change and its ramifications, this report focuses on developing a representative set of measurable metrics that are likely to be affected by climate change over the next 20-25 years and that, when taken together, can be used as indicators of environmental sustainability.

U.S. Environmental Protection Agency. Office of Atmospheric Programs. 2012. Climate Change Indicators in the United States, 2012. 2nd Edition.

http://www.epa.gov/climatechange/science/indicators

This report from the U.S. Environmental Protection Agency presents 26 indicators in five categories: greenhouse gases, weather and climate, oceans, snow and ice, and society and ecosystems. It includes a discussion of each indicator and data trends, including the indicator's limitations and the data sources(s). An online version of this report will be updated periodically as new data become available.

Examples of Single Sector Indicators

AGRICULTURAL INDICATORS

Tubiello, F.N. and C. Rosenzweig. 2008. Developing climate change impact metrics for agriculture. *The Integrated Assessment Journal* 8 (1): 165-84. $

http://journals.sfu.ca/int_assess/index.php/iaj/article/view/276/240

This journal article proposes a framework for analyzing the benefits of climate change adaptation policies on the agricultural sector, identifying biophysical factors, agricultural system characteristics, socioeconomic data, and climate policy as key categories for analysis. Based on this framework, a set of metrics is proposed to help policy and decision makers evaluate, quantify, and communicate the benefits of climate change policies on agricultural systems. It also discusses improvements needed in current agronomic and economic models to address key uncertainties in assessing benefits of climate policies.

ECOLOGICAL INDICATORS

Brody, S.D., W.G. Peacock, and J. Gunn. 2012. Ecological indicators of flood risk along the Gulf of Mexico. *Ecological Indicators* 18: 493-500. $

http://www.sciencedirect.com/science/article/pii/S1470160X12000192

This journal article reports on the measurement and analysis of ecological indicators of flood risk along the Gulf of Mexico. The researchers identified and measured four ecological indicators critical to protecting communities from the adverse impacts of floods (floodplain area, soil porosity, naturally occurring wetlands, and pervious surfaces) and statistically tested the degree to which they reduce flood losses observed across the study area over a five-year period.

Harley, M. and J. van Minnen. 2009. Development of Adaptation Indicators. European Topic Centre on Air and Climate Change.

http://acm.eionet.europa.eu/reports/ETCACC_TP_2009_6 ETCACC TP 2009 6 Adapt Ind

This technical paper presents a theoretical and practical framework for the development of adaptation indicators, differentiating between process- and outcome-based indicators. It discusses application of the framework in two case studies, development of adaptation indicators for the biodiversity sector and regional adaptation strategies and indicators, and includes potential indicators for each.

Harley, M. and J. van Minnen. 2010. Adaptation Indicators for Biodiversity. European Topic Centre on Air and Climate Change.
http://acm.eionet.europa.eu/reports/ETCACC_TP_2010_15_Adap_Ind_Biodiv

This technical paper discusses efforts to develop adaptation indicators for biodiversity and reduce the vulnerability of the European Union to climate change. Among these efforts is an October 2010 expert workshop that resulted in an approach to the development of adaptation indicators for biodiversity based on seven adaptation principles. The report summarizes the adaptation principles, related measures and possible indicators, and each measure's relevance to the biodiversity and associated sectors (agriculture, energy and transport, fresh water, forestry, land use planning, marine and fisheries).

Score, A., R.M. Gregg, and L.J. Hansen. 2012. Monitoring Climate Effects in Temperate Marine Ecosystems: A Test-Case Using California's MPAs. MPA Monitoring Enterprise, California Ocean Science Trust.
http://monitoringenterprise.org/pdf/Monitoring_climate_change_effects_in_temperate_marine_ecosystems.pdf

This technical report recommends a framework for incorporating consideration of climate change into marine protected area (MPA) monitoring efforts to increase the effectiveness of adaptive MPA management. The three-tiered framework provides scalable implementation options for managers that can track climate change impacts and provide "alerting signals" for marine ecosystems.

Smithers, R., N. Kent, K. Miller, J. van Minnen, and M. Harley. 2011. Climate Change Adaptation Indicators for Biodiversity. European Topic Centre on Air and Climate Change Mitigation.
http://acm.eionet.europa.eu/reports/ETCACM_TP_2011_14_CCadapt_ind_biodiv

This technical paper reports on a project for the European Environmental Agency that builds on earlier work to develop climate change adaptation indicators. The project used the high-level biodiversity adaptation indicator categories proposed in *Adaptation Indicators for Biodiversity* (see p. 24) as a starting point for discussions with stakeholders from the European Commission and member states in the biodiversity, agriculture, forestry, and water policy areas. The paper summarizes the results of a stakeholder survey and interviews. Four respondents noted existing suites of climate change adaptation indicators for their policy area. The paper also provides brief summaries of relevant ongoing policy initiatives at the European Union level that may have a direct bearing on the development, synthesis, and embedding of biodiversity adaptation indicators.

The Heinz Center. 2008. The State of the Nation's Ecosystems 2008: Measuring the Lands, Waters, and Living Resources of the United States—Highlights.
http://www.heinzctr.org/content/state-nation%E2%80%99s-ecosystems-2008-0

This highlights report is derived from a larger report of the same name available from Island Press ($ http://islandpress.org/ip/books/book/islandpress/S/bo8013860.html). It is the result of an effort to identify key aspects (indicators) of the nation's six principal ecosystems—coasts and oceans, fresh waters, forests, farmlands, grasslands and shrublands, and urban and suburban areas—so that they can be tracked through time to provide a consistent and comprehensive view of trends in each of these ecosystems and the nation as a whole. The 108 indicators identified are grouped into four categories: extent and pattern indicators; chemical and physical characteristics; biological components indicators; and goods and services indicators.

U.S. Global Change Research Program. 2011. Ecosystem Responses to Climate Change: Selecting Indicators and Integrating Observation Networks. NCA Report Series, Volume 5a.
http://library.globalchange.gov/national-climate-assessment-ecological-indicators-workshop-report

This report summarizes the discussions and outcomes of a workshop on developing indicators for ecosystem responses to climate change and integrating observational networks, which was conducted to support the development of the 2013 National Climate Assessment (NCA). The goals of the workshop were to outline a process for selecting indicators that represent the impacts of climate change on the nation's ecosystems and identify opportunities for collaboration and coordination among existing and potential future observational networks that could be used to improve the understanding of these impacts. Background materials and presentations are available at https://sites.google.com/a/usgcrp.gov/eco-monitoring/

ECOLOGICAL INDICATORS

The Montréal Process. n.d. The Montréal Process Sustainable Forest Management Criteria and Indicators. Website. Accessed July 15, 2013.
http://www.montrealprocess.org/Resources/Criteria_and_Indicators

The Montréal Process is a voluntary international effort by 12 member countries, including the United States, to support sustainable forest management. One of the first steps in this effort was the development of seven criteria that characterize the essential components of sustainable forest management and indicators to describe, monitor, assess, and report on national forest trends and progress toward sustainable forest management. Working group members reported their indicator results around 2003 and again around 2009. This website includes a booklet that describes the process and the current 54 indicators, the indicators for the 2003 and 2009 reports, **country** reports, and technical reports.

HEALTH INDICATORS

English, P.B., A.H. Sinclair, Z. Ross, H. Anderson, V. Boothe, C. Davis, K. Ebi, B. Kagey, K. Malecki, R. Shultz, and E. Simms. 2009. Environmental health indicators of climate change for the United States: Findings from the State Environmental Health Indicator Collaborative. *Environmental Health Perspectives* 117 (11): 1673-81. $
http://www.jstor.org/stable/40382450

This journal article reports on efforts of the State Environmental Health Indicators Collaborative to develop a recommended list of environmental health indicators for climate change. The list, based on a scientific literature review to identify climate change outcomes and actions, includes environmental, population vulnerability, mitigation, adaptation, and policy indicators of climate change. The researchers concluded that data exist for many environmental and health measures but that more evaluation of their sensitivity and usefulness is needed and further attention is necessary to increase data quality and availability and to develop new surveillance databases.

PHYSICAL CLIMATE INDICATORS

Scotland and Northern Ireland Forum for Environmental Research. 2006. An Online Handbook of Climate Trends across Scotland.
http://www.climatetrendshandbook.adaptationscotland.org.uk

This handbook presents the changes in climate across Scotland in the last century and provides a benchmark against which future climate change can be measured. Twenty-four variables are divided into four categories: temperature-related variables, precipitation-related variables, air pressure-related variables, and sunshine-related variables. For each group of variables, the handbook presents trend data, a table summarizing average change over the period for which data is available by region, a map showing changes over time for the whole of Scotland, and an analysis of the data.

U.S. Global Change Research Program. 2011. Monitoring Climate Change and Its Impacts: Physical Climate Indicators. NCA Report Series, Volume 5b.
http://library.globalchange.gov/national-climate-assessment-physical-climate-indicators-workshop-report

This report summarizes the discussions and outcomes of a workshop focused on monitoring changes in the physical climate system to support the development of the 2013 National Climate Assessment (NCA). The goal of the workshop was to identify a few broad categories of potential physical climate indicators using a set of priorities developed by the NCA and to provide a clear justification for how they would inform the nation about climate change. Additional goals included providing input on the overall NCA framework for selecting the indicators and suggesting methodologies to construct indicators. Background materials and presentations are available at https://sites.google.com/a/usgcrp.gov/physical-indicators

RESEARCH INDICATORS

National Research Council. Committee on Metrics for Global Change Research. 2005. Thinking Strategically: The Appropriate Use of Metrics for the Climate Change Science Program. The National Academies Press.
http://www.nap.edu/catalog.php?record_id=11292

At the request of the U.S. Climate Change Science Program (CCSP), the National Research Council established a committee to develop quantitative metrics and performance measures for documenting progress and evaluation of future performance of the research program. This committee report describes approaches that industry, academia, and federal agencies have taken to measure research performance; lays out principles for developing metrics; discusses uncertainty reduction; and proposes a set of general metrics for assessing the progress of CCSP program elements and for guiding future strategic planning.

SOCIETAL INDICATORS

California Environmental Protection Agency. Office of Environmental Health Hazard Assessment. 2010. Indicators of Climate Change in California: Environmental Justice Impacts.
http://oehha.ca.gov/multimedia/epic/epic123110.html

This report presents four indicators to track how certain socioeconomic or racial groups in California may be experiencing disproportionately greater impacts on their health or well-being than others as a result of climate change. The report includes the selection criteria for the indicators, and for each indicator there is a description, an analysis of data, a discussion of why the indicator is important and the factors that influence the indicator, and information on the data characteristics (including strengths and limits).

Cutter, S.L., C.G. Burton, and C.T. Emrich. 2010. Disaster resilience indicators for benchmarking baseline conditions. *Journal of Homeland Security and Emergency Management* 7 (1): 1-22. $
http://dx.doi.org/10.2202/1547-7355.1732

This journal article provides a methodology and a set of indicators to serve as the baseline set of conditions from which to measure the effectiveness of programs, policies, and interventions designed to improve disaster resilience. It includes the results of an application of the methodology to the Southeastern United States.

Droesch, A.C., N. Gaseb, P. Kurukulasuriya, A. Mershon, K.M. Moussa, D. Rankine, and A. Santos. 2008. A Guide to Vulnerability Reduction Assessment. United Nations Development Programme.
http://betterevaluation.org/sites/default/files/Attachment CBA VRA Guide Dec 08.pdf

This guide from the United Nations Development Programme (UNDP) describes the Vulnerability Reduction Assessment (VRA) approach, which is an element of UNDP's monitoring and evaluation framework for climate change adaptation projects. The VRA is designed to measure the changing climate vulnerabilities of communities and to be comparable across vastly different projects, regions, and contexts, making it possible to determine if a given project is successful or unsuccessful in reducing climate change risks. It comprises four indicators that are based on context-specific, perception-based questions, the answers to which are collected through repeated evaluations of community perceptions of project effectiveness and climate change risks.

U.S. Global Change Research Program. 2011. Climate Change Impacts and Responses: Societal Indicators for the National Climate Assessment. NCA Report Series, Volume 5c.
http://library.globalchange.gov/national-climate-assessment-societal-indicators-workshop-report

This report summarizes the presentations, discussions, and outcomes of a workshop focused on developing societal indicators to support the development of the 2013 National Climate Assessment (NCA). The workshop participants provided input on a number of topics, including categories of societal indicators for the NCA, alternative approaches to constructing indicators and the better approaches for NCA to consider, specific requirements and criteria for implementing the indicators, and sources of data for, and creators of, such indicators. The report also includes a white paper about developing societal indicators for the NCA, an extensive societal indicators bibliography, and a societal indicator inventory.

Wongbusarakum, S. and C. Loper. 2011. Indicators to Assess Community-Level Social Vulnerability to Climate Change: An Addendum to SocMon and SEM-Pasifika Regional Socioeconomic Monitoring Guidelines. Global Socioeconomic Monitoring Initiative for Coastal Management.
http://www.socmon.org/download.ashx?docid=64623

This document provides a set of socioeconomic indicators related to climate change that can be included in a socioeconomic assessment to understand community-level climate change vulnerability and to inform coastal management needs and adaptive management. It is an addendum to regional socioeconomic monitoring guidelines from the Global Socioeconomic Monitoring Initiative for Coastal Management (SocMon) and SEM-Pasifika and is a first draft for field testing, circulation, and revision.

URBAN CLIMATE CHANGE INDICATORS

Jacob, K. and R. Blake. 2010. Indicators and monitoring. In *Climate Change Adaptation in New York City: Building a Risk Management Response. Annals of the New York Academy of Sciences* 1196: 127-41. http://onlinelibrary.wiley.com/doi/10.1111/j.1749-6632.2009.05321.x/abstract

As part of a larger report from the New York City Panel on Climate Change, this chapter examines potential urban climate change indicators for New York City that could support the implementation of flexible adaptation pathways for the region's critical infrastructure. It discusses indicators related to changes in the climate, climate science, climate impacts, and adaptation activities that can be devised and tracked over time and illustrates a proposed structure for a climate change indicators and monitoring process or system. The Adaptation Assessment Guidebook (the report's Appendix B) provides additional guidance on monitoring and reassessing climate change indicators and is available at: http://onlinelibrary.wiley.com/doi/10.1111/j.1749-6632.2010.05324.x/abstract

Examples of Report Cards and Performance Scorecards

Committee on Climate Change (United Kingdom). 2013. Adaptation Indicators. Accessed August 23, 2013. http://www.theccc.org.uk/charts-data/adaptation-indicators

This website provides report cards with trend information for indicators in six sectors: coastal, flooding, water scarcity, agriculture and forestry, wildlife, and upland peat. The indicators are drawn from the report *Managing the Land in a Changing Climate-Adaptation Sub-Committee Progress Report 2013*, which is available at http://www.theccc.org.uk/publication/managing-the-land-in-a-changing-climate

Commonwealth Scientific and Industrial Research Organisation. 2012. Marine Climate Change in Australia: Impacts and Adaptation Responses 2012 Report Card. http://www.oceanclimatechange.org.au/content/index.php/2012/home

This report card summarizes the current knowledge of climate change impacts to Australia's marine environment and marine species, what is likely to happen in the future, and what Australia is doing to address the impacts. The report card is accompanied by full topic reports that include a scientific review of existing data and literature and information on confidence assessments, adaptation responses, observation and modeling, and references.

Living with Environmental Change. 2013. Terrestrial Biodiversity Climate Change Impacts: Report Card 2012-2013. http://www.lwec.org.uk/resources/report-cards/biodiversity

This report card highlights key trends affecting biodiversity in the United Kingdom. For each trend there is a summary of the latest scientific evidence along with confidence levels and a description of potential future impacts. The report card is based on 15 technical reports that provide detailed supporting evidence. Updates are planned every two years.

Marine Climate Change Impacts Partnership. n.d. Annual Report Card and Briefing Notes. http://www.mccip.org.uk/annual-report-card.aspx

This report card provides an annual account of developments in marine climate science in the United Kingdom to help inform climate change-related decision-making. It addresses the current state of scientific understanding of marine climate change and the associated level of confidence for what is already happening and what could happen in the future in four categories: climate of the marine environment, healthy and diverse marine ecosystem, clean and safe seas, and commercially productive seas. Annual summaries are accompanied by full topic reports that include detailed supporting evidence and sections on knowledge gaps, social and economic impacts, and confidence assessments.

Marine Climate Change Impacts Partnership. 2009. Marine Climate Change Impacts: Exploring Ecosystem Linkages.
http://www.mccip.org.uk/ecosystem-linkages.aspx

In addition to its annual comprehensive report cards (see previous entry), the Marine Climate Change Impacts Partnership produces special topic reports that focus on key topics in more detail. This ecosystems report card looked at five key issues (CO_2 and ocean acidification, Arctic sea-ice loss, seabirds and food webs, nonnative species, and coastal economies) to show how the interconnected nature of the marine ecosystem magnifies the many discrete impacts of climate change documented in the annual report cards.

Marine Climate Change Impacts Partnership. 2013. Marine Climate Change Impacts: Fish, Fisheries & Aquaculture.
http://www.mccip.org.uk

This report card focuses on how climate change is affecting fish and shellfish in the United Kingdom and what the social and economic consequences could be. Findings are summarized under four themes: changes in species distributions, implications for marine management, social and economic consequences, and the wider (global) picture. Maps show regional stories about what is happening now and what could happen in the future.

Sempier, T.T., D.L. Swann, R. Emmer, S.H. Sempier, and M. Schneider. 2010. Coastal Resilience Index: A Community Self-Assessment. MASGP-08-014. Mississippi-Alabama Sea Grant Consortium.
http://masgc.org/page.asp?id=591

The Coastal Resilience Index is a tool communities can use to examine how prepared they are for storms and storm recovery. It provides a simple, inexpensive method for community leaders to perform a self-assessment of their community's resilience to coastal hazards, identifies weaknesses a community may want to address prior to the next hazard event, and guides discussion within a community. After completing the self-assessment, a resilience index (low, medium, or high) can be calculated that serves as an indicator of a community's ability to reach and maintain an acceptable level of functioning and structure after a disaster.

U.S. Environmental Protection Agency. Office of Water. 2013. 2012 Highlights of Progress Responses to Climate Change by the National Water Program.
http://water.epa.gov/scitech/climatechange/2012-National-Water-Program-Strategy.cfm

This is the U.S. Environmental Protection Agency's first climate change progress report for implementation of the *National Water Program 2012 Strategy: Response to Climate Change*. The strategy focuses on five long-term programmatic vision areas: water infrastructure, watersheds and wetlands, coastal and ocean waters, water quality, and working with tribes. The report includes a scorecard for tracking progress based on the seven stages or phases (initiation, assessment, response development, initial implementation, robust implementation, mainstreaming, and monitoring of outcomes and adaptive management) of development of climate response programs. Programs receive a numerical score based on their phase of development.

U.S. Environmental Protection Agency. Office of Water. 2012. National Coastal Condition Report IV.
http://water.epa.gov/type/oceb/assessmonitor/nccr

This report provides both a national and regional scorecards for the state of the nation's coastal waters. The scorecards are based on indices for water quality, sediment quality, benthic, coastal habitat, and fish tissue contaminants. The report includes an analysis of the data and information on the indices, sources of data, and ranking scales.

U.S. Forest Service. Office of the Climate Change Advisor. 2011. A Performance Scorecard for Implementing the U.S. Forest Service Climate Change Strategy 2010-2015.
http://www.fs.fed.us/climatechange/advisor/scorecard.html

The U.S. Forest Service developed this scorecard (and *Navigating the Climate Change Performance Scorecard*, a guidance document) for National Forests and Grasslands to track progress in addressing climate change between 2011 and 2015. The scorecard addresses four dimensions: organizational capacity, engagement, adaptation, and mitigation. By 2015, each unit is expected to answer yes to at least 7 of the 10 scorecard questions, with at least one yes in each dimension. The goal is to create a balanced approach to climate change that includes managing forests and grasslands to adapt to changing conditions, mitigating climate change, building partnerships across boundaries, and preparing employees to understand and apply emerging science.

Case Study Databases

A number of websites provide access to resources and case studies related to climate adaptation actions. Two of these websites include a number of case studies and resources related to monitoring and evaluation of climate adaptation actions.

EcoAdapt and Island Press. n.d. Climate Adaptation Knowledge Exchange (CAKE). Accessed July 15, 2013.
http://www.cakex.org

The Climate Adaptation Knowledge Exchange (CAKE) is aimed at building a community of practice and shared knowledge base for managing natural systems in the face of rapid climate change. It features a directory of practitioners to share knowledge and strategies, data tools and information available from other sites, and a virtual library. Resources include documents and case studies related to monitoring and evaluation.

Global Environment Facility Evaluation Office. n.d. Climate-Eval. Accessed April 10, 2013.
http://www.climate-eval.org

The Climate-Eval website is a venue to share information on the evaluation of climate change and development interventions. It includes hundreds of climate-related evaluations and a few of those are specific to climate adaptation.

Evaluation Organizations

There are a number of organizations dedicated to evaluation. The organizations in this section have information and resources on their websites, and their members may be sources of help for those interested in monitoring and evaluation of climate adaptation actions.

Action Research Community for Adaptation in Bangladesh
http://www.arcab.org

The Action Research Community for Adaptation in Bangladesh (ARCAB) is a long-term research program, jointly managed by the International Institute for Environment and Development and Bangladesh Centre for Advanced Studies, focused on looking at the effectiveness of community-based adaptation (CBA) in Bangladesh. ARCAB will be researching the effectiveness of CBA and is committed to communicating the results to inform policy, practice, and future research in Bangladesh and globally.

American Evaluation Association
http://www.eval.org

The American Evaluation Association (AEA) is an international professional association of evaluators devoted to the application and exploration of program evaluation, personnel evaluation, technology, and many other forms of evaluation. AEA has approximately 7,300 members representing over 60 countries.

Conservation Measures Partnership
http://www.conservationmeasures.org

The Conservation Measures Partnership (CMP) is a partnership of conservation organizations that seek better ways to design, manage, and measure the impacts of their conservation actions. CMP members work together on issues related to impact assessment and accountability because they believe that collectively they have a greater chance of designing and implementing effective monitoring and evaluation systems and enhancing program and project design and implementation. CMP strives to promote innovation in monitoring and evaluation of conservation efforts and serves as a catalyst within the conservation community.

Environmental Evaluators Network
http://www.environmentalevaluators.net

The Environmental Evaluators Network (EEN) aims to advance the field of environmental evaluation through more systematic and collective learning, including fostering connections among evaluators and evaluation consumers, growing the body of knowledge of environmental program and policy evaluation, and fostering the collaborative development of products for advancing the field. EEN comprises environmental, conservation, and natural resource evaluators and evaluation consumers from academia, consulting organizations, foundations, government agencies, and nonprofit organizations.

International Initiative for Impact Evaluations
http://www.3ieimpact.org

The International Initiative for Impact Evaluations (3ie) is an organization devoted to improving lives through better policies, programs, and projects designed and implemented based on evidence drawn from high-quality impact evaluations. 3ie supports the production of new evidence by financing new policy-relevant impact evaluations and supporting the synthesis of existing evidence on particular topics. It also provides technical support to organizations wanting to carry out their own impact evaluations; engages with policy makers to make them aware of the importance of impact evaluation, encouraging them to commission and learn from impact studies; and supports efforts to build capacity in low- and middle-income countries to carry out impact evaluations. 3ie efforts to promote the production and use of quality impact evaluations include a website and publications.

SEA Change
http://www.seachangecop.org

SEA Change is a community of practice focused on developing a culture of high quality and rigorous monitoring and evaluating frameworks, approaches, and methodologies for adaptive responses to climate change interventions and practices in Southeast Asia. The organization focuses on fostering partnerships and building member capacity; sharing best practices, lessons learned, guidelines, approaches, methods, tools, and innovations; and influencing policy and practice around monitoring and evaluation of adaptive responses to climate change interventions.

Glossary

The following are standard definitions for commonly used monitoring and evaluation terms.

Note: The field of monitoring and evaluation does not yet have universally accepted definitions, so the resources cited in this bibliography may each define these terms differently.

Checklist—A list of factors, properties, aspects, components, criteria, tasks, or dimensions to be considered in order to perform a certain task as part of an evaluation.

Developmental Evaluation—An evaluation approach that is based on a long-term partnership between evaluators and those engaged in innovative initiatives and development and that supports adaptive learning.

Evaluation—A systematic collection and study of information to assess how well a program or project is working and why. An evaluation answers specific questions about performance and may focus on assessing operations or results. Results may be used to assess effectiveness, identify how to improve performance, and/or guide resource allocation.

Impact Evaluation—An evaluation that systematically and empirically investigates the long-term impacts produced by a program or project.

Implementation Evaluation (also called Process Evaluation)—An evaluation designed to address the quality or efficiency of program operations or their fidelity to program design. An implementation evaluation is often conducted in the beginning stages of a new program or initiative to provide managers with quick feedback on whether action is needed to help get the program up and running as intended.

Indicator—A quantifiable unit of measure that measures conditions over time. The term indicator is often used to describe both a performance measure and a measurement that a program has no specific control over.

Logic Model—A depiction, usually graphic, of how a program expects to reach its goal(s) and includes the relationships between inputs (resources, including human, financial, and organizational), activities, outputs (direct products of program activities), and both short- and long-term outcomes.

Metric—A standard of measurement that is composed of both a performance measure and a target.

Mixed Method Evaluation—An evaluation that combines both quantitative and qualitative techniques to answer one or more evaluation questions.

Monitoring—The systematic and routine collection of information related to the implementation of programs and/or projects. The information is often used for accountability, tracking progress toward goals, learning, and informing decisions.

Performance Measurement—A systematic, ongoing collection, analysis, and reporting of information regarding performance to determine progress toward pre-established goals or standards.

Performance Measure—A quantifiable unit of measure that conveys something about the type or level of program activity conducted (process), the direct product or service delivered by a program (output), and/or the results of that product or service (outcome).

Performance Scorecard—A communication tool used to track and provide information on progress toward priority performance measures. A scorecard often uses graphics to quickly convey information.

Report Card—A communication tool used to summarize and provide trend information for conditions of interest (e.g., environmental conditions). A report card often uses graphics to quickly convey information and is often supplemented with additional data and information in supporting documents.

Target—A quantifiable goal for performance measures that programs aim to achieve, usually within a set period of time.

www.ingramcontent.com/pod-product-compliance
Lightning Source LLC
Chambersburg PA
CBHW081413170526
45166CB00010B/3329